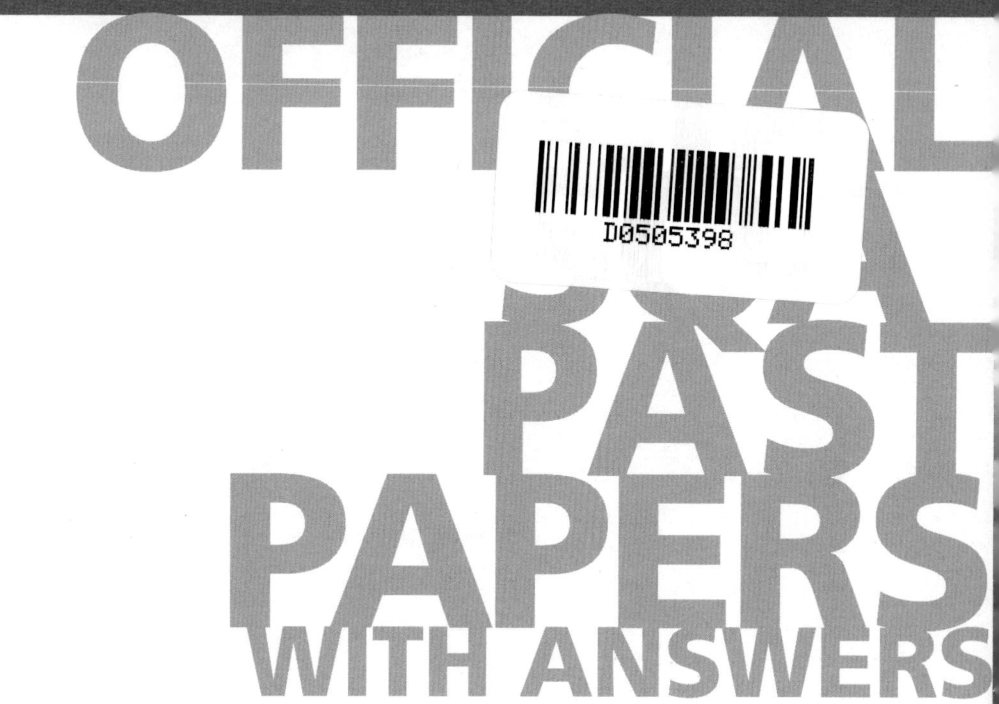

HIGHER

PRODUCT DESIGN
2010-2013

HODDER GIBSON
LEARN MORE

Hodder Gibson is grateful to the copyright holders, as credited on the final page of the Question Section, for permission to use their material. Every effort has been made to trace the copyright holders and to obtain their permission for the use of copyright material. Hodder Gibson will be happy to receive information allowing us to rectify any error or omission in future editions.

Hachette UK's policy is to use papers that are natural, renewable and recyclable products and made from wood grown in sustainable forests. The logging and manufacturing processes are expected to conform to the environmental regulations of the country of origin.

Orders: please contact Bookpoint Ltd, 130 Park Drive, Abingdon, Oxon OX14 4SE. Telephone: (44) 01235 827720. Fax: (44) 01235 400454.

Lines are open 9.00–5.00, Monday to Saturday, with a 24-hour message answering service. Visit our website at www.hoddereducation.co.uk. Hodder Gibson can be contacted direct on: Tel: 0141 848 1609; Fax: 0141 889 6315; email: hoddergibson@hodder.co.uk

This collection first published in 2013 by

Hodder Gibson, an imprint of Hodder Education,

An Hachette UK Company

2a Christie Street

Paisley PA1 1NB

BrightRED PUBLISHING Hodder Gibson is grateful to Bright Red Publishing Ltd for collaborative work in preparation of this book and all SQA Past Paper and National 5 Model Paper titles 2013.

Typeset by PDQ Digital Media Solutions Ltd, Bungay, Suffolk NR35 1BY

Printed in the UK

A catalogue record for this title is available from the British Library

ISBN 978-1-4718-0299-7

3 2 1

2014 2013

Introduction

Study Skills – what you need to know to pass exams!

Pause for thought

Many students might skip quickly through a page like this. After all, we all know how to revise. Do you really though?

Think about this:

"IF YOU ALWAYS DO WHAT YOU ALWAYS DO, YOU WILL ALWAYS GET WHAT YOU HAVE ALWAYS GOT."

Do you like the grades you get? Do you want to do better? If you get full marks in your assessment, then that's great! Change nothing! This section is just to help you get that little bit better than you already are.

There are two main parts to the advice on offer here. The first part highlights fairly obvious things but which are also very important. The second part makes suggestions about revision that you might not have thought about but which WILL help you.

Part 1

DOH! It's so obvious but ...

Start revising in good time

Don't leave it until the last minute – this will make you panic.

Make a revision timetable that sets out work time AND play time.

Sleep and eat!

Obvious really, and very helpful. Avoid arguments or stressful things too – even games that wind you up. You need to be fit, awake and focused!

Know your place!

Make sure you know exactly **WHEN and WHERE** your exams are.

Know your enemy!

Make sure you know what to expect in the exam.

How is the paper structured?

How much time is there for each question?

What types of question are involved?

Which topics seem to come up time and time again?

Which topics are your strongest and which are your weakest?

Are all topics compulsory or are there choices?

Learn by DOING!

There is no substitute for past papers and practice papers – they are simply essential! Tackling this collection of papers and answers is exactly the right thing to be doing as your exams approach.

Part 2

People learn in different ways. Some like low light, some bright. Some like early morning, some like evening / night. Some prefer warm, some prefer cold. But everyone uses their BRAIN and the brain works when it is active. Passive learning – sitting gazing at notes – is the most INEFFICIENT way to learn anything. Below you will find tips and ideas for making your revision more effective and maybe even more enjoyable. What follows gets your brain active, and active learning works!

Activity 1 – Stop and review

Step 1

When you have done no more than 5 minutes of revision reading STOP!

Step 2

Write a heading in your own words which sums up the topic you have been revising.

Step 3

Write a summary of what you have revised in no more than two sentences. Don't fool yourself by saying, 'I know it but I cannot put it into words'. That just means you don't know it well enough. If you cannot write your summary, revise that section again, knowing that you must write a summary at the end of it. Many of you will have notebooks full of blue/black ink writing. Many of the pages will not be especially attractive or memorable so try to liven them up a bit with colour as you are reviewing and rewriting. **This is a great memory aid, and memory is the most important thing.**

Activity 2 — Use technology!

Why should everything be written down? Have you thought about 'mental' maps, diagrams, cartoons and colour to help you learn? And rather than write down notes, why not record your revision material?

What about having a text message revision session with friends? Keep in touch with them to find out how and what they are revising and share ideas and questions.

Why not make a video diary where you tell the camera what you are doing, what you think you have learned and what you still have to do? No one has to see or hear it but the process of having to organise your thoughts in a formal way to explain something is a very important learning practice.

Be sure to make use of electronic files. You could begin to summarise your class notes. Your typing might be slow but it will get faster and the typed notes will be easier to read than the scribbles in your class notes. Try to add different fonts and colours to make your work stand out. You can easily Google relevant pictures, cartoons and diagrams which you can copy and paste to make your work more attractive and **MEMORABLE**.

Activity 3 – This is it. Do this and you will know lots!

Step 1

In this task you must be very honest with yourself! Find the SQA syllabus for your subject (www.sqa.org.uk). Look at how it is broken down into main topics called MANDATORY knowledge. That means stuff you MUST know.

Step 2

BEFORE you do ANY revision on this topic, write a list of everything that you already know about the subject. It might be quite a long list but you only need to write it once. It shows you all the information that is already in your long-term memory so you know what parts you do not need to revise!

Step 3

Pick a chapter or section from your book or revision notes. Choose a fairly large section or a whole chapter to get the most out of this activity.

With a buddy, use Skype, Facetime, Twitter or any other communication you have, to play the game "If this is the answer, what is the question?". For example, if you are revising Geography and the answer you provide is "meander", your buddy would have to make up a question like "What is the word that describes a feature of a river where it flows slowly and bends often from side to side?".

Make up 10 "answers" based on the content of the chapter or section you are using. Give this to your buddy to solve while you solve theirs.

Step 4

Construct a wordsearch of at least 10 X 10 squares. You can make it as big as you like but keep it realistic. Work together with a group of friends. Many apps allow you to make wordsearch puzzles online. The words and phrases can go in any direction and phrases can be split. Your puzzle must only contain facts linked to the topic you are revising. Your task is to find 10 bits of information to hide in your puzzle but you must not repeat information that you used in Step 3. DO NOT show where the words are. Fill up empty squares with random letters. Remember to keep a note of where your answers are hidden but do not show your friends. When you have a complete puzzle, exchange it with a friend to solve each other's puzzle.

Step 5

Now make up 10 questions (not "answers" this time) based on the same chapter used in the previous two tasks. Again, you must find NEW information that you have not yet used. Now it's getting hard to find that new information! Again, give your questions to a friend to answer.

Step 6

As you have been doing the puzzles, your brain has been actively searching for new information. Now write a NEW LIST that contains only the new information you have discovered when doing the puzzles. Your new list is the one to look at repeatedly for short bursts over the next few days. Try to remember more and more of it without looking at it. After a few days, you should be able to add words from your second list to your first list as you increase the information in your long-term memory.

FINALLY! Be inspired...

Make a list of different revision ideas and beside each one write **THINGS I HAVE** tried, **THINGS I WILL** try and **THINGS I MIGHT** try. Don't be scared of trying something new.

And remember – "FAIL TO PREPARE AND PREPARE TO FAIL!"

Higher Product Design

The course

The Higher Product Design course consists of three units – Product Evaluation, Developing Design Proposals and Manufacturing. Together these units give candidates experience in what is required to make products commercially viable.

How the course is assessed

The grade you get for Higher Product Design depends upon three things:

- The internal assessments you do in school or college ("NAB's") – while these don't count towards your final grade, you must pass them before you can achieve an award at Higher.
- Your Design Assignment – this is usually issued by the SQA in late January and can be at any time from then during the spring term. It is submitted in April and is worth 70 marks from the course assessment of 140 marks and is assessed by the SQA.
- The written paper – this is sat during the main diet of exams and is worth 70 marks.

The Design Assignment

The Design Assignment consists of four briefs based upon a scenario. It is very important to read and fully understand the briefs before starting. They are designed to allow candidates with a wide range of skills to have access to marks. Some candidates may be very creative aesthetically, while others may have a more technical background. Some briefs, therefore, may allow candidates with an aesthetic flair to be creative, while others are more technical in nature. It is very important to choose carefully which brief suits you so that you can perform to your best.

Candidates usually perform very well in the ideas generation part of the Design Assignment (15 marks), but only the most able candidates perform well in the development part of the Design Assignment (30 marks). Many candidates develop two proposals and then go through an evaluation process to come to a proposal. This is fine as long as there is no duplication in the work. Candidates who only develop one proposal tend to look at both aesthetic development and technical development to perform. Candidates who do well also include examples of valid modelling. There must be a reason to model, conclusions from the tests and information on how this has affected the design. This also pertains to any use of research material – how has this information affected and been used in the development of a design proposal?

To gain full marks in the development section of the Design Assignment, you must look at both technical and aesthetic issues.

The rest of the marks are awarded for Communication (10 marks), Recording and Justification of Decisions (10 marks) and the Final Proposal (5 marks).

Communication

Marks are awarded across the whole folio. Examiners are looking for flow, clarity and a variety of graphic techniques. They will also assess whether any modelling element is valid and in context.

Recording and justification

Again, this is marked across the whole folio. Examiners are looking for recording of decisions being made, reasoning for modifications, evaluations of material choices, identification of manufacturing methods, reasoning based upon performance, properties of materials, scale of manufacture, aesthetics, target market etc.

Design proposal

This is based upon the final graphic presented for the proposal. Including related additional supportive text may increase your mark slightly.

Exam

The first question of the exam is a standard question worth 30 marks. While this question does not vary very much in its layout, the products do change from year to year. Candidates cannot assume that answers which were acceptable in previous exams will still be appropriate because materials and manufacturing processes will change. Generic answers should be avoided because answers which are not clearly in context will not accrue marks. It is the candidate's responsibility to show knowledge and understanding of issues, not the marker's responsibility to interpret answers.

The product graphics and technical information in this question are given to contextualise the subject matter. Any lifting of this information will accrue no marks.

Question 1

Question 1 (a) concerns the product specification and has the same opening question every year, although with different products to be considered. Try to put yourself in the scenario of going to visit a designer with a wish list of requirements for the product. This is a design specification which would give the designer scope to be creative as possible. It is important that you do not take technical information from the question stem.

The next two questions, 1 (b) and 1 (c), concern materials and production processes. Materials identified and justified must be applicable to the product and the properties must tie in with the materials used. Generic answers such as 'durable', 'strong' and 'lightweight' are not sufficient unless in context. Likewise, production methods must be in context for the component, the use, the material, the volume of production etc.

The topics being examined may vary in the rest of question 1. They depend upon the products being used as subject matter. Questions based upon the appeal to the target market, health and safety in relation to both manufacturer and consumer, and ergonomics have all been asked in the past. All answers are expected to be in context to the subject matter.

At Higher level, candidates will not be asked to describe a method of production but could be asked to justify why a production method has been chosen, or to choose an appropriate production method and justify an answer in the context of a certain product or component. The understanding is in the appreciation that there could be several methods of production for components but that in certain circumstances there is an optimal method. This is how differentiation is achieved.

How to answer questions

In general, only responses for question 1 (a) can be answered with bullet points. If a question starts with 'State' then a one word or one line answer is acceptable. However, at Higher level, most questions will ask for an explanation, a justification or a description. These types of questions require extended responses and the candidate must show understanding of the subject matter – they should try to make their points as clear and unambiguous as possible. In general, this cannot be done adequately with a one or two line bullet point answer. Where bullets are used, there is a danger that the marker will think the answer is not expanded enough to gain either full or partial marks. In some cases, even when the candidate is on the right track or shows some knowledge, their response is not complete so does not justify full marks.

Candidates often use published marking schemes as a basis for answering practice papers, but marking schemes are not answer scripts and are used by markers to access discussion points that are expected to turn up in responses. Each of these points would accrue a mark if included in a response correctly and in context to the question.

The graphics in the question papers are sometimes there to enhance the questions, but sometimes to help with questions on aesthetics or ergonomics. They should be used to trigger points that the candidates might go on to make.

In general, if a question is worth 5 marks in an extended question, then the examiner is expecting five pieces of fully justified, explained or described information, which is valid and in context to the subject matter.

General advice

Work through past papers as much as possible and answer questions as you would expect to answer your final exam paper (based upon the advice above).

Give yourself a time limit for each practice paper. Make it as real as possible!

Only check the marking scheme when you have finished a paper. Retry the questions you felt you did badly.

Throughout the course, try to look at products and identify the materials used and the manufacturing processes involved – think about why certain approaches have been taken.

The rewards for passing Higher Product Design could be the start of a fulfilling career in design or in manufacturing. Be confident and trust your instincts. Most of the answers are common sense when you give some thought to the problem or issue.

GOOD LUCK!

HIGHER

2010

X211/301

NATIONAL QUALIFICATIONS 2010

TUESDAY, 18 MAY
1.00 PM – 3.00 PM

PRODUCT DESIGN HIGHER

70 marks are allocated to this paper.

Attempt ALL questions.

SECTION A

1. Two bathroom accessory ranges are shown below.

Accessory Range A

Comprises waste bin, toilet brush and holder, toothbrush holder, tumbler and soap dish.

Manufactured from frosted acrylic.

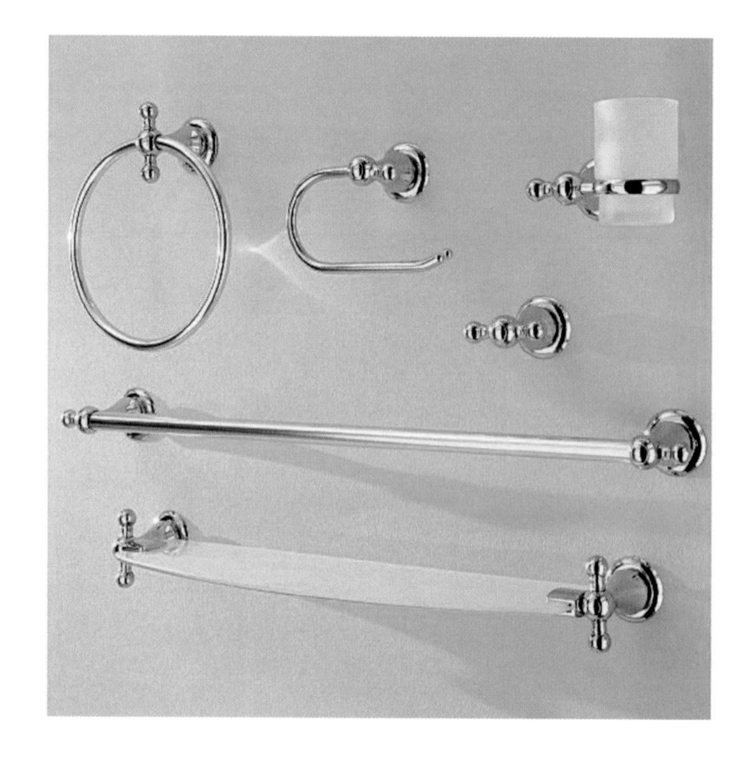

Accessory Range B

Comprises toilet roll holder, hand towel ring, tumbler ring including tumbler, robe hook and towel rail.

Manufactured from chrome plated metal and glass.

Glass shelf sold separately (available in plain or frosted glass).

Marks

1. (continued)

Each of the bathroom accessory ranges shown have been designed for well-known high street retailers.

(*a*) Write a product specification for **one** of the bathroom accessory ranges in relation to its target market. **6**

(*b*) Justify the choice of materials used to produce **both** ranges. **6**

(*c*) Justify the production processes that would be suitable for the manufacture of **both** ranges of bathroom accessories. **6**

(*d*) Explain how **both** bathroom accessory ranges have been influenced by function. **4**

(*e*) Describe the appeal of **both** bathroom accessory ranges from the consumer's viewpoint. **4**

(*f*) Describe the ergonomic issues associated with these types of products. **4**

Total for Section A (30)

[Turn over

Marks

SECTION B

2. **Annotated sketches**, **working drawings** and **rendered 3D computer models** are three techniques used to communicate ideas during the design process.

For each of the techniques above:

Identify a stage in the design process where it could be used and explain why it is appropriate to that stage. **(6)**

3. Developments in manufactured boards and fittings have enabled designers to create a wide range of affordable flat-pack furniture styles.

(*a*) **Describe** the environmental issues surrounding the development of man-made boards. 2

(*b*) **Explain** why "flat-packed" furniture is popular with:

(i) consumers; 2

(ii) manufacturers. 2

(*c*) **Explain** how manufacturers ensure that flat-pack products can be easily assembled by consumers. 2

High quality furniture, using solid timbers and traditional construction methods, is still available to the consumer.

(*d*) **Explain** why there is still a market for this type of furniture. 2

(10)

Marks

4. The components shown below have been manufactured by the process of compression moulding.

Smoke alarm casing

Car tailgate

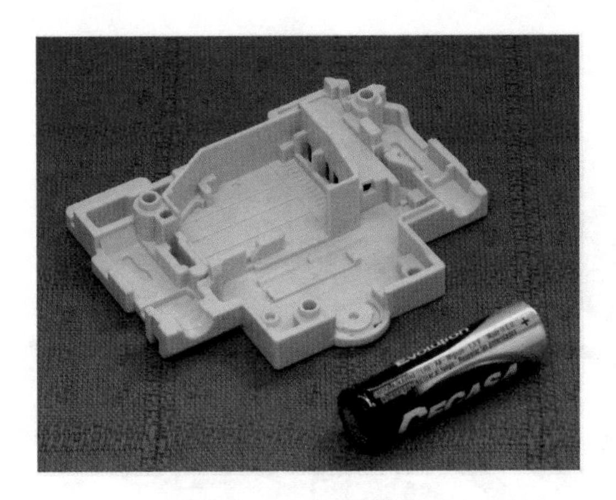

Battery housing

(*a*) **Explain** why compression moulding is suitable for these types of product. **2**

(*b*) **State** the name of a suitable material and explain why it is appropriate for the manufacture of components. **3**

(5)

[Turn over

Marks

5. Research is important in ensuring the success of a product.

(*a*) **Describe** the research information that would be required for the stages below.

(i) Initial concept — 2

(ii) Planning for production — 2

(*b*) **Describe** the information that would be identified by end user trials. — 2

(6)

6. The base and frame of the car jack shown below has been manufactured by the processes of piercing and blanking, then press forming.

(*a*) **Explain** why these processes are suitable for manufacturing the components for this product. — 3

Standard components such as nuts, bolts and rivets have been used in the car jack frame.

(*b*) **Explain** why it is appropriate to assemble the frame using these types of components. — 2

(*c*) **Describe** the advantages to the manufacturer in using standard components. — 2

(7)

Marks

7. A designer has been asked by a well-known manufacturer to produce designs for a range of electronic personal organisers.

(*a*) **Describe** how advances in technology have influenced products such as the electronic device shown above. **2**

(*b*) **Explain** the implicatons of IPR (Intellectual Property Rights) associated with this new range of products for:

(i) the designer; **2**

(ii) the client/manufacturer. **2**

(6)

Total for Section B (40)

[*END OF QUESTION PAPER*]

[BLANK PAGE]

HIGHER

2011

X211/301

NATIONAL QUALIFICATIONS 2011

TUESDAY, 17 MAY
1.00 PM – 3.00 PM

PRODUCT DESIGN HIGHER

70 marks are allocated to this paper.

Attempt ALL questions.

SECTION A

1.

Chair A—Ergolife Stol

Made from solid birch and fabric.

Rolled up length: 400 mm

Diameter: 100 mm

Weight approx 1 kg

Retail price £29·99

Chair A

Chair B

Chair B—Messenger Bag Director's Chair

Less than 100 mm thick when folded.

Comes with telescoping legs and folding backrest.

Made from aluminium, plastic and nylon mesh seat.

Integrated cup holder and two pockets in the armrests.

Retail price £95

Marks

1. (continued)

(*a*) Outline a product specification for the design of **one** of the chairs. **6**

(*b*) Justify the choice of materials used to produce **each** of these chairs. **6**

(*c*) Identify **and** justify the production methods chosen to manufacture **one** of the chairs. **6**

(*d*) For both chairs describe the quality assurance issues that would affect:

(i) the manufacture;

(ii) the consumer. **4**

(*e*) For **each** of the chairs shown, identify a market niche where their use would be appropriate and justify your answer. **4**

(*f*) Describe the ergonomic issues associated with the use of **each** of these chairs. **4**

Total for Section A (30)

[Turn over

Marks

SECTION B

2. The teeth in the blade of the cheese grater shown have been manufactured using the process of piercing and blanking.

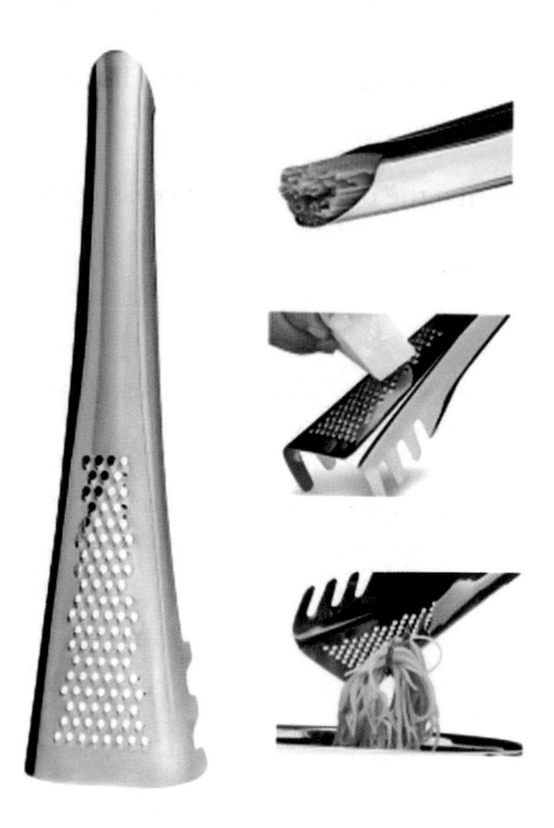

(*a*) **Explain** why piercing and blanking is a suitable process for this type of product. **2**

(*b*) **State** the name of the process used to form the product. **1**

(*c*) **State** the name of a suitable metal for the manufacture of this product **and** justify why it is appropriate. **3**

(6)

Marks

3. An in-house designer has been asked by Bosch to produce a new product concept for their lawnmower range.

(*a*) **Explain** why the designer may prefer an "**open brief**" rather than a "**closed brief**". **2**

(*b*) **Describe** the information a designer would gain from an initial meeting with the client. **2**

A designer leaves the company during the design of the product.

(*c*) **Describe** the "intellectual property rights" issues relating to the design of the product in this situation. **3**

(*d*) **Describe two** methods that companies can use to protect intellectual property. **2**

The rapid prototype model shown in Figure 1 has been produced for testing purposes.

(*e*) **State** the name of a suitable rapid prototyping process to address the key features to be tested for the product in Figure 1. **1**

(10)

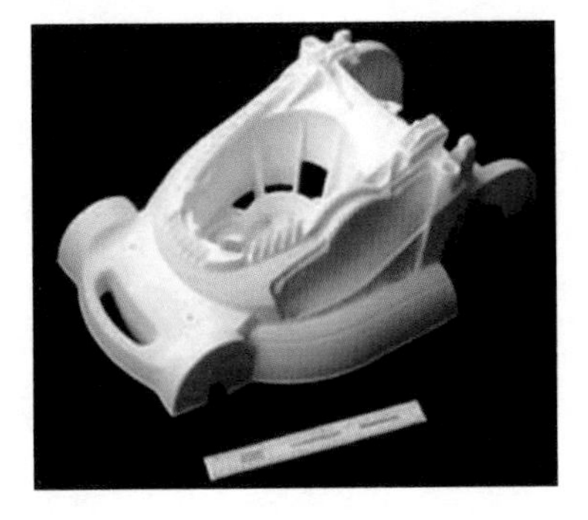

Figure 1

Key Features to be tested:
Form and fit testing
Durability and functional material testing
Environmental testing
High stress, heat and chemical resistance testing

[Turn over

Marks

4. Two kayaks are shown below.

Figure 2 is produced by a rotational moulded polyethylene.

Figure 3 is produced by traditional methods using glass reinforced plastic.

Rotational moulded Kayak

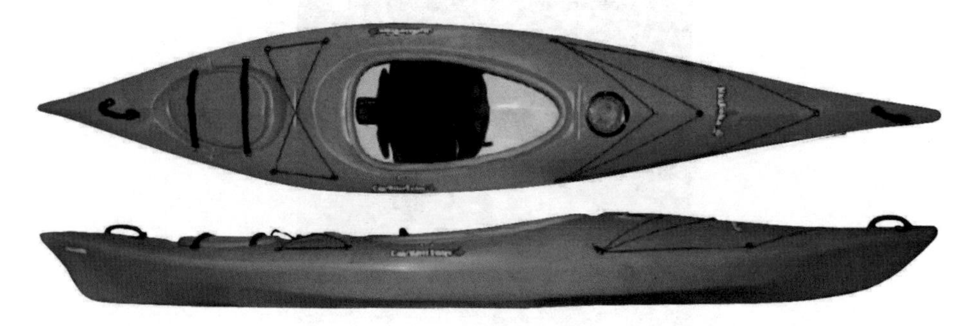

Figure 2

Glass reinforced plastic Kayak

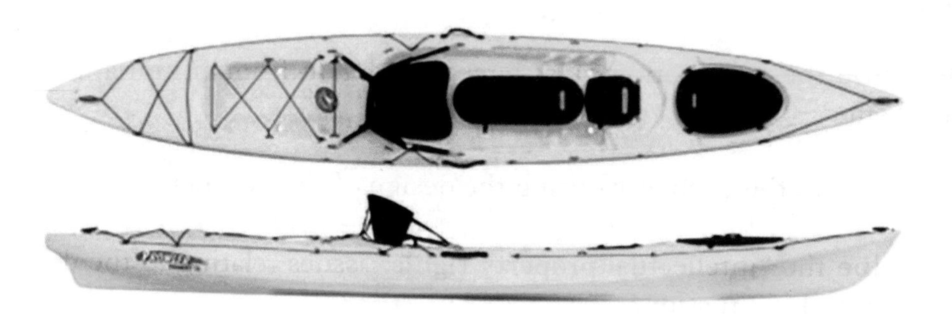

Figure 3

(*a*) **Explain** why each method of production is suitable for a product of this type. **4**

(*b*) Identify **two** manufacturing disadvantages of using rotational moulding for this product. **2**

(6)

Marks

5. The new Glasgow Museum of Transport is due to open this year on the banks of the River Clyde.

"The historical development of the Clyde is a unique legacy. The building is a tunnel-like shed, which is open at opposite ends to the city and the Clyde. It creates a journey from the external into the world of the exhibits. The museum positions itself as open and fluid with its engagement of context and content. The building is conceived as an extrusion open at opposing ends."

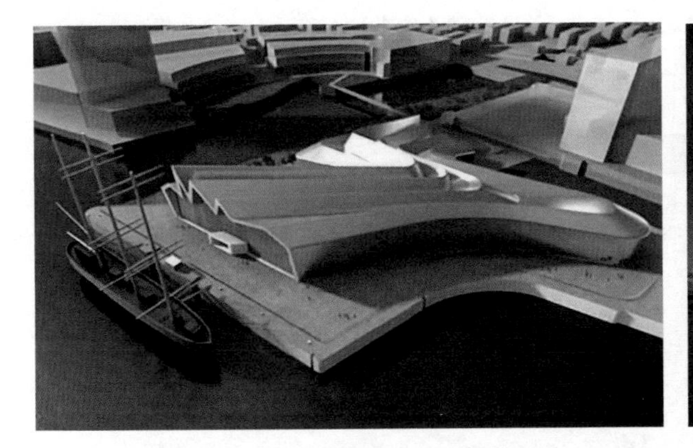

Describe how aesthetics have influenced this design. **4**

(4)

[Turn over

Marks

6. A GANTT chart, similar to the one shown below, was created by the production engineer prior to the manufacture of a product.

Production Planning Chart

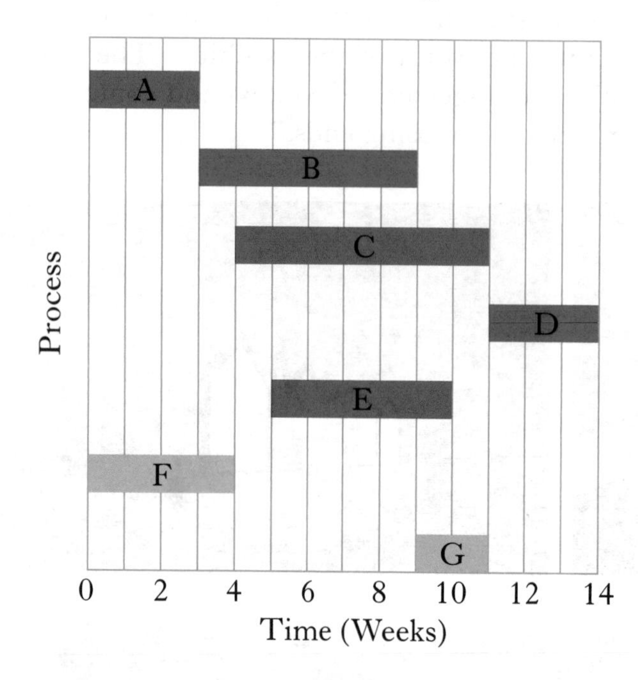

A Tool setting
B Moulding
C Fabrication
D Fitting
E Machining
F Sourcing components
G Component delivery

(*a*) **Describe** the advantages of the GANTT chart in the production planning process. **3**

Some of the component parts of the product may be manufactured by sub-contractors.

(*b*) **Explain two** disadvantages of using sub-contractors. **2**

(*c*) **Explain** what steps the manufacturer could take to minimise this problem. **2**

(7)

Marks

7. Two sledges are shown below.

Sledge A—BMW Sauber F1 Sled

Made from a composite plastic

Sledge B

Traditional wooden sledge made from solid birch with metal runners

(*a*) **Explain** and **justify** the advantages of the composite material used for sledge A over the traditional materials used in sledge B. **4**

(*b*) **Describe** how "form follows function" has influenced the design of sledge A. **3**

(7)

Total for Section B (40)

[*END OF QUESTION PAPER*]

HIGHER

2012

X211/12/01

NATIONAL QUALIFICATIONS 2012

THURSDAY, 24 MAY
1.00 PM – 3.00 PM

PRODUCT DESIGN HIGHER

70 marks are allocated to this paper.

PB X211/12/01 6/8060

Attempt all questions

SECTION A

1. The products shown below have been designed and manufactured for use in pre-school, early years and primary schools.

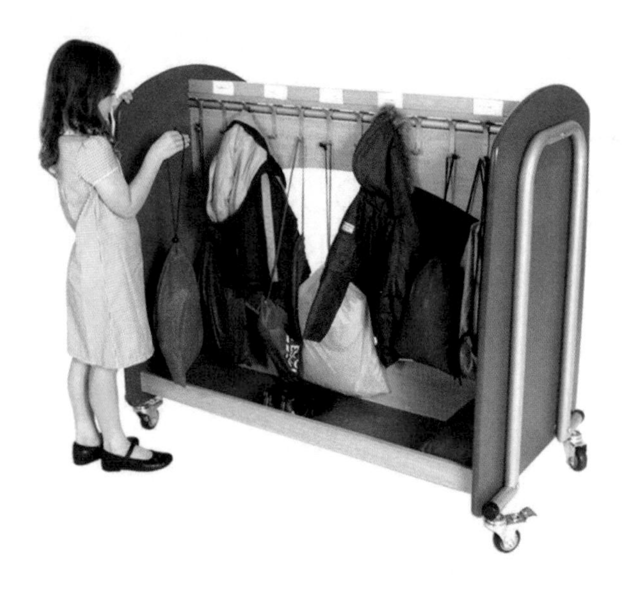

Mobile Cloakroom Trolley

Laminated MDF Panels

Scratch resistant mild steel tubular frame

12 mild steel double hooks

Large base storage area

Self-assembly required

Retail price £235

Cubby Storage Unit

Strengthened tubular PVC plastic frame

4 HDPE plastic bins for shoe and bag storage

Wall mounted

Retail price £70

1. (continued) *Marks*

(*a*) Outline a product specification for the design of **one** of the coat racks. **6**

(*b*) Justify the choice of materials used to produce **both** of these coat racks. **6**

(*c*) Justify the production methods that could be chosen to manufacture **both** coat racks. **6**

(*d*) For **both** coat racks describe the safety issues that would affect:

(i) the manufacture;

(ii) the consumer. **4**

(*e*) Explain how the design of **both** of these coat racks has been influenced by functional issues. **4**

(*f*) Explain the issues surrounding the choice of **both** of the coat racks for:

(i) the target market;

(ii) end user. **4**

Total for Section A (30)

[Turn over

SECTION B

Marks

2. The garden kneeler shown below allows the user to work in comfort.

The product has been manufactured using the process of blow moulding.

(*a*) **State** the name of a suitable material for the manufacture of the kneeler and justify your choice. **2**

(*b*) State **two** features that would indicate that the kneeler was manufactured using the process of blow moulding. **2**

(*c*) Justify why blow moulding was used for the manufacture of the kneeler. **1**

(5)

3. Modelling is an important part of the design process.

For each of the modelling types below, describe the information that would be gathered from their use.

The three modelling types are:

- **Scale models** **2**
- **Test models** **2**
- **Prototypes**. **2**

(6)

Marks

4. Many companies such as Dyson register their name as a trademark.

(*a*) Explain why they do this. **2**

A company such as Dyson will create many new designs like the Ball™ shown below.

They will then patent these designs.

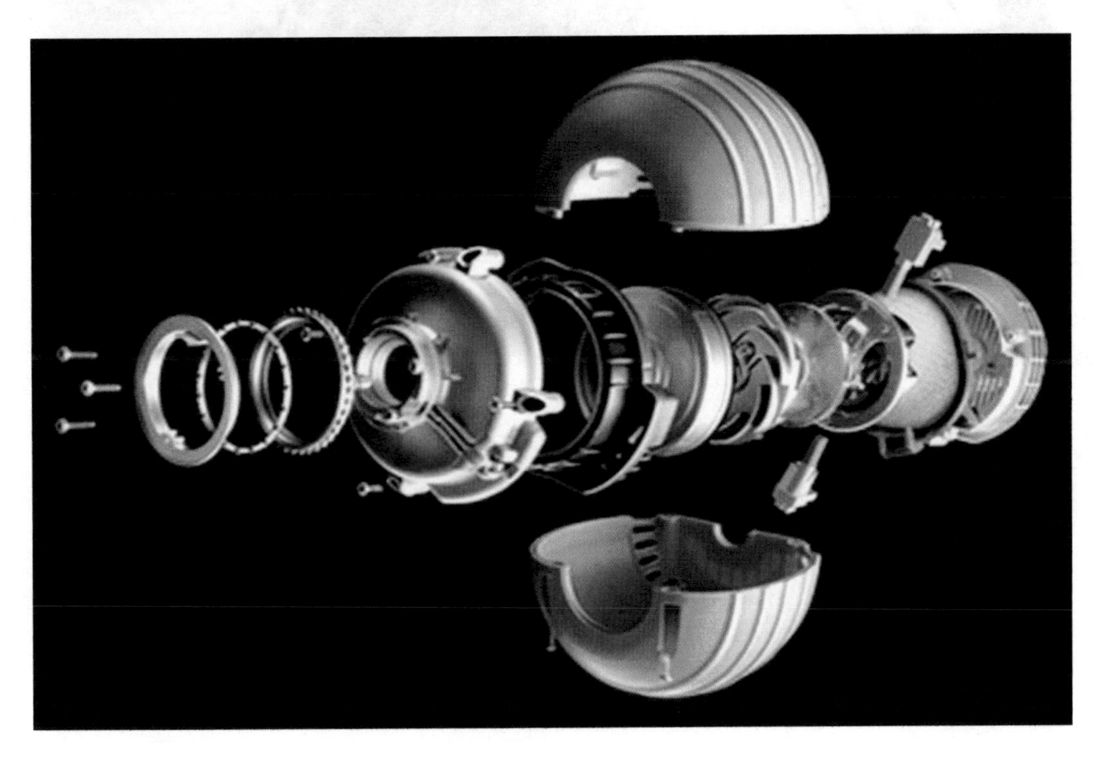

Ball™ technology

(*b*) Explain why companies patent their designs. **2**

(4)

[Turn over

Marks

5. The Evac+Chair is used to transport a mobility impaired individual down stairs in a safe, smooth, and controlled way.

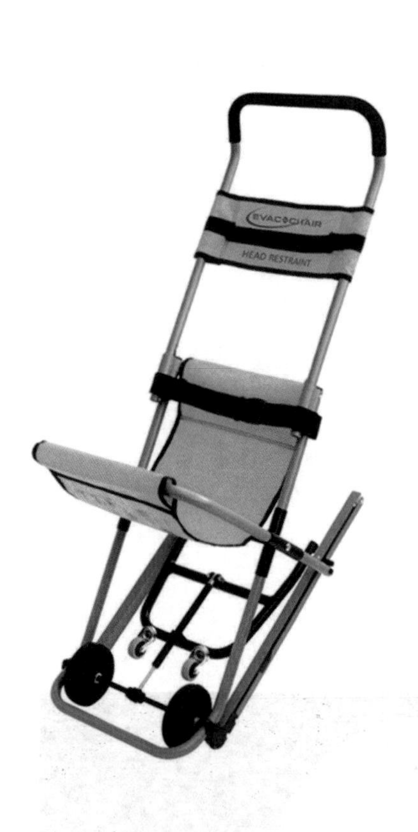

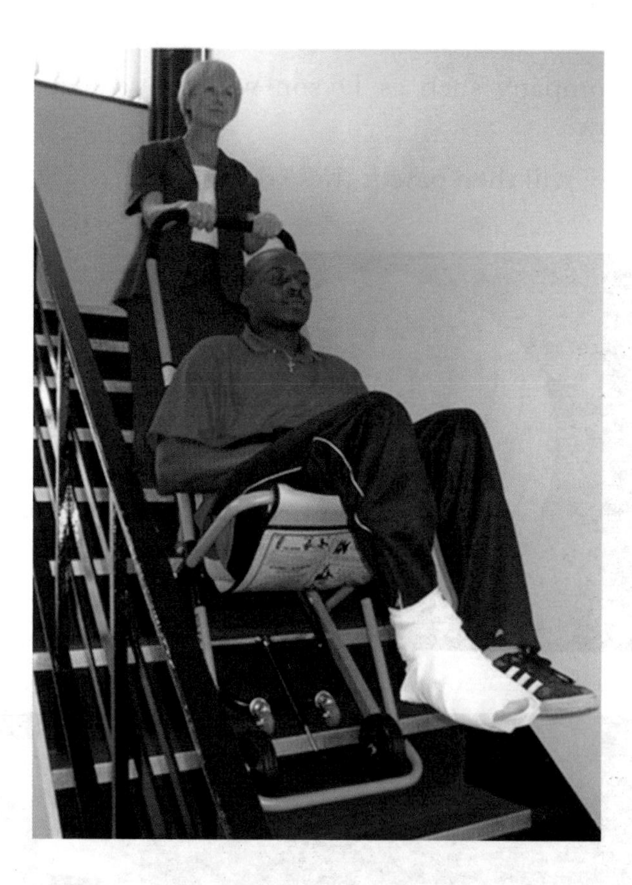

The consideration of ergonomic issues in the design of this product was vitally important.

With reference to the Evac+Chair, describe the ergonomic issues that may have been considered in terms of:

(*a*) Anthropometrics; **3**

(*b*) Physiology; **3**

(*c*) Psychology. **3**

(9)

Marks

6. The car wheel shown below is a one-piece aluminium alloy sand casting.

It has been chrome plated.

(*a*) Justify the choice of sand casting for the manufacture of this wheel. **2**

(*b*) Explain why additional machining was necessary after sand casting the wheel. **2**

A product such as the alloy wheel shown above can also be manufactured by Pressure Die Casting.

(*c*) Explain the benefits gained by manufacturing the wheel using this process. **3**

(*d*) Explain the benefits of using alloys rather than pure metals. **2**

(9)

[Turn over for Question 7 on *Page eight*

Marks

7. Luxottica Group S.p.A. is a manufacturer of high quality sunglasses.

Made of gold-plated metal with mineral glass lenses. Originally designed for US military fighter pilots in 1937.

The Ray-Ban® Aviator

Made from carbon fibre for legs and a resin composite lens. Designed as part of the 2009 range.

The Ray-Ban® Tech

In the design of a product such as sunglasses, certain factors have a major influence on the outcome of the product.

(*a*) Other than Aesthetics and Ergonomics, **justify two** other factors that you consider to be important in the design of these products. **4**

Companies such as Luxottica must keep up to date with their designs to ensure their continued success within the market place.

Fashion and **Style** are two important factors that ensure that Luxottica remain a world leader in the manufacture of sunglasses.

(*b*) Explain the difference between **Fashion** and **Style** in relation to these products. **3**

(7)

Total for Section B (40)

[*END OF QUESTION PAPER*]

HIGHER

2013

X211/12/01

NATIONAL QUALIFICATIONS 2013

WEDNESDAY, 29 MAY
1.00 PM – 3.00 PM

PRODUCT DESIGN HIGHER

70 marks are allocated to this paper.

PB X211/12/01 6/8060

Attempt all questions

SECTION A

1. Each of the lawnmowers shown below have been designed for a well known high street DIY retailer.

Electric Hover Mower

Blade—Mild Steel
Handle—Plastic coated Mild Steel
Body—Polypropylene
Gears/Fasteners—Nylon
Cable Length—20 Metres

Weight— 4·2 kg

Retail price £29·99

Cylinder Mower (Manually Operated)

Blade—HSS (Tool Steel)
Handle—Foam Rubber coated Aluminium
Body—Mild Steel
Gears/Fasteners—Nylon
Wheels—Metal Alloy
Grass Catcher—Nylon with Polypropylene base

Weight—7·3 kg

Retail price £119·50

Marks

1. (continued)

(*a*) Write a product specification for **one** of the lawnmowers in relation to its target market. **6**

(*b*) Justify the choice of materials used to produce **both** lawnmowers. **6**

(*c*) Identify and justify the production processes that could be used to manufacture **both** lawnmowers. **6**

(*d*) Explain the ergonomic issues associated with **both** lawnmowers. **4**

(*e*) Describe the appeal of **both** lawnmowers from the consumer's viewpoint. **4**

(*f*) Describe how the design of **both** lawnmowers has been influenced by functional issues. **4**

Total for Section A (30)

[Turn over

Marks

SECTION B

2. The body of the adjustable spanner shown below is made by the process of drop forging.

(*a*) Explain why drop forging is a suitable process for producing the body of this adjustable spanner. **1**

(*b*) State **two** features that would indicate that this product was made by drop forging. **2**

(*c*) State a suitable material that could be used for the body of the spanner and give a reason for your choice. **2**

(5)

Marks

3. Aesthetics is a major consideration in the design of a product such as the Sky+ remote control shown below.

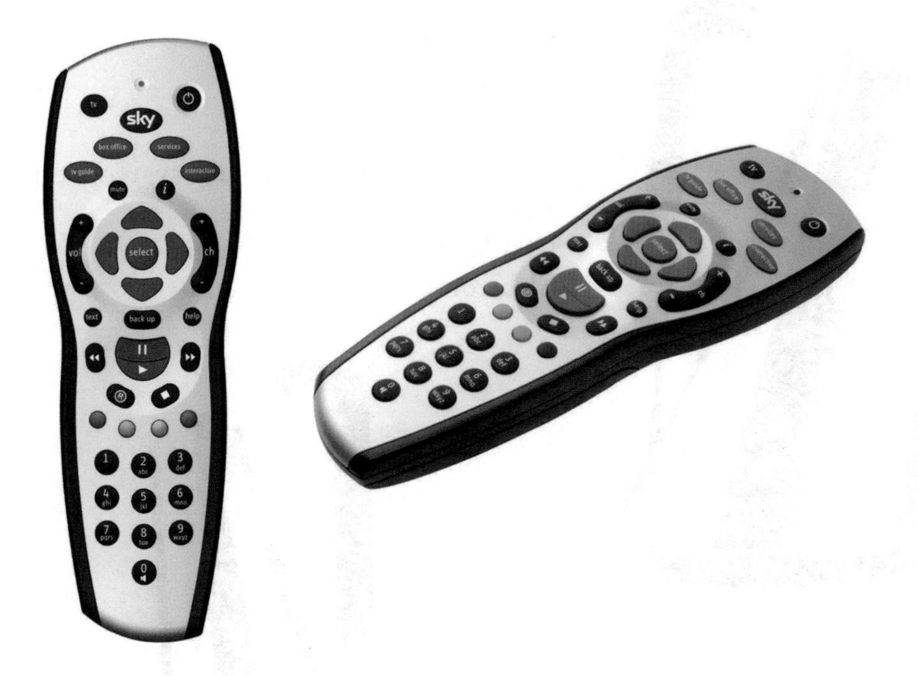

Describe where **four** aspects of aesthetics have influenced the design of the Sky+ remote control. **4**

(4)

[Turn over

Marks

4. A designer has been asked to produce concepts for a new style of domestic kettle.

Specification

Stainless steel body
Programmable timers
Protective thermal security system stops overheating
Save up to 25% more energy
Display integrated in the handle
Electronic temperature control

Bugatti Vera Electric Kettle—£189·95

The kettle shown above has been designed for a niche market.

(*a*) With reference to the kettle explain the term "**niche market**". **2**

Another selling point is that the kettle could be recycled easily.

(*b*) Describe the steps the designer could take to make the kettle easier to recycle at the end of its working life. **2**

The kettle could be manufactured using batch production techniques.

(*c*) Describe the considerations the manufacturer would need to make before deciding upon this production system. **2**

(6)

Marks

5. The graph shown below has been used to predict and compare how well **two** new graphics tablets will sell.

Product Life Cycle (Sales and Profit)

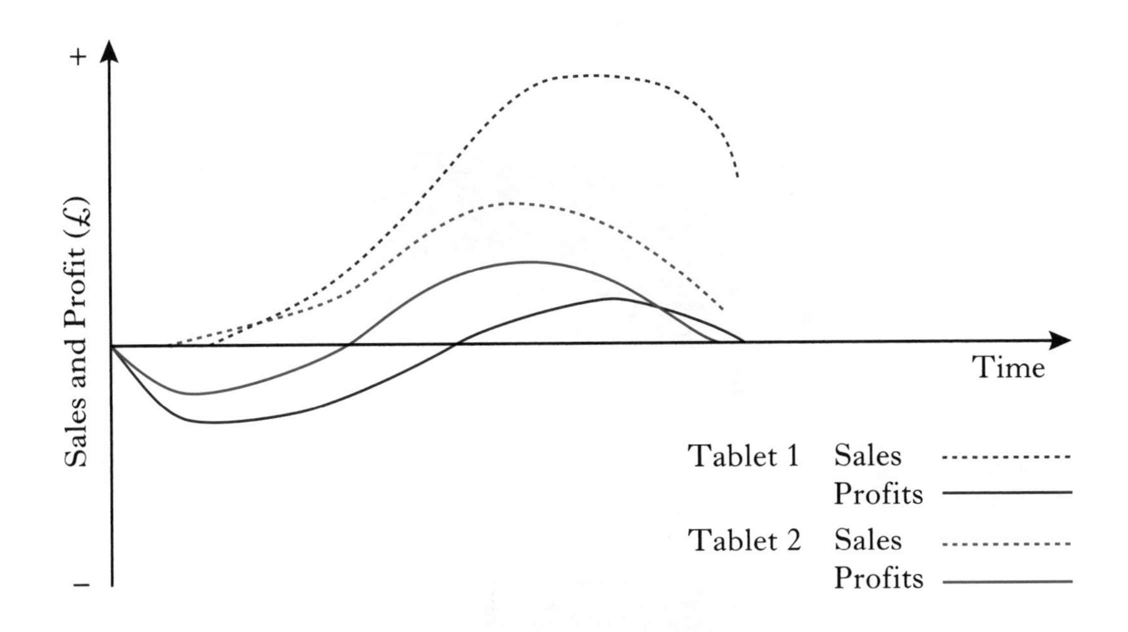

(*a*) Describe what steps a manufacturer could take to reduce the time required to introduce a product onto the market. **2**

(*b*) From the graph above state which of the two graphics tablets would be commercially viable and explain your reasons for this choice. **3**

(*c*) Describe how a company could extend the sales life of a product. **2**

(7)

[Turn over

Marks

6. The carcass of the kitchen cabinet shown below has been constructed using manufactured boards and knock down fittings.

(*a*) Explain the benefits to the **manufacturer** of using knock down fittings instead of traditional joining methods. **2**

The door of the kitchen cabinet is manufactured using solid timber.

(*b*) Explain the benefits to the **consumer** of using solid timber for the cabinet doors. **2**

(*c*) Describe the obsolescence issues associated with modern fitted kitchens. **2**

(6)

Marks

7. During design development many designers use CAD software to simulate the behaviour of products.

(*a*) Explain the benefits of computer simulation over user trials with prototype models. **2**

(*b*) A prototype model of a car disc brake was produced using Fused Deposition Modelling.

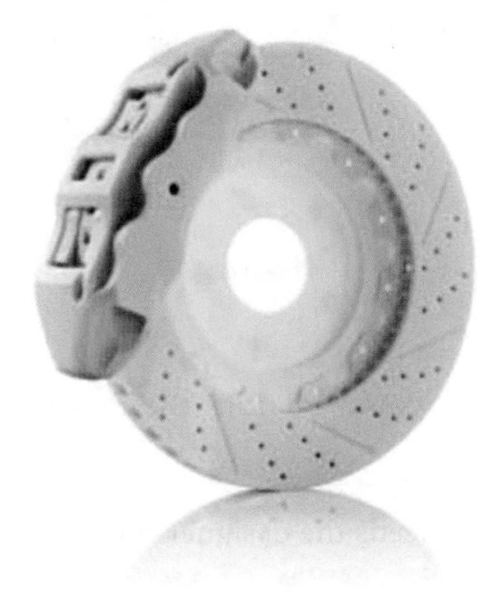

(i) Name a suitable material that could be used for the Fused Deposition Modelling process. **1**

(ii) State **one** advantage and **one** disadvantage associated with Fused Deposition Modelling. **2**

(5)

[Turn over for Question 8 on *Page ten*

Marks

8. A company has commissioned a designer to produce a range of kitchen accessories suitable for users with limited manual dexterity.

(*a*) Explain how the designer could identify the needs of the user group before developing concept ideas. **1**

(*b*) Describe the physiological needs the designer might find within the user group. **3**

(*c*) Describe **two** idea generation techniques that could be used to help produce concept ideas. **2**

(*d*) Describe a technique that could be used to present the design concepts to the client. **1**

(7)

Total for Section B (40)

[*END OF QUESTION PAPER*]

[BLANK PAGE]

Acknowledgements

Permission has been sought from all relevant copyright holders and Hodder Gibson is grateful for the use of the following:

A photograph of a Peugeot Elis Electric Peppermill and Saltmill reproduced with permission of Peugeot (2009 page 2);

A photograph of a Dyson DC24 Bagless Upright Vacuum Cleaner © Dyson (2009 page 5);

A photograph of an Eames La Chaise © 1946 Eames Office, LLC (2009 page 8);

A photograph of a lounge chair designed by P Hvidt & O Molgaard-Nielsen, 1947 © Hvidt & Mølgaard A/S (2009 page 8);

A photograph of a Cegasa battery © Cegasa Internacional (2010 page 5);

A photograph of a BlackBerry® Bold™ 9700. BlackBerry® and related trademarks, names and logos are the property of BlackBerry Limited and are registered and/or used in the U.S. and countries around the world. Used under license from BlackBerry Limited (2010 page 7);

Two photographs of Ergolife Stol chairs © Clever Little Ideas (2011 page 2);

A photograph of a Messenger Bag Director's Chair © Hammacher Schlemmer (2011 page 2);

A photograph of a Bosch lawnmower © Bosch Lawn & Garden Limited (2011 page 5);

A photograph of a prototype model © 3T RPD Ltd (2011 page 5);

Photographs of the Glasgow Museum of Transport. Reproduced with permission of Zaha Hadid Architects (2011 page 7);

A photograph of a BMW Sauber F1 Sled © BMW (2011 page 9);

Two photographs of a Hozelock garden kneeler © Hozelock UK Ltd (2012 page 4);

Two photographs showing Dyson Ball™ technology © Dyson (2012 page 5);

Two photographs © Evac+Chair International Ltd (2012 page 6);

Two photographs of Ray-Ban® Aviator and Ray-Ban® New Tech sunglasses © Luxottica Group (2012 page 8);

Image © samodelkin8/Shutterstock.com (2013 page 4);

Two photographs of SKY remote controls © British Sky Broadcasting Group plc (2013 page 5);

A photogaph of a Bugatti Vera Electric Kettle © Ilcar di Bugatti S.r.l. (2013 page 6);

Image © Christopher David Howells/Shutterstock.com (2013 page 8).

HIGHER | ANSWER SECTION

SQA HIGHER PRODUCT DESIGN 2010–2013

HIGHER PRODUCT DESIGN 2010

SECTION A

1. (a) The accessory ranges must:
   - Be stable and sturdy
   - Hold the intended items
   - Be manufactured from durable materials that are suitable and appropriate for the Bathroom accessory's function
   - Be safe (complying to appropriate health & hygiene standards)
   - Be priced to suit the intended target market
   - Ensure aesthetics suit the market niche or consumer aspirations
   - Be easily installed by target market (set B only)
   - Production costs significantly less than selling price
   - Be produced in a variety of colours (Set A) or plated finishes and styles (Set B) to give target market a wider choice
   - Be easy to use
   - Be easy to clean
   - Not corrode
   - Replaceable brush heads
   - Able to be mass produced
   - Any other suitable statement.
   - Lightweight/easily moved.
   - Recyclable.
   - Fit a modern bathroom.

   Six statements at one mark each

   (b) Statements which identify issues such as:
   - Durability of material [*non corrosion/scratch resistant (Set B only)*]
   - Strength to weight issues
   - Safety
   - Suitability for production methods
   - Function of component parts
   - Aesthetic properties
   - Ease of clean/hygiene
   - Re-cycling
   - Form of material eg decorative ends (set B) are suitable for mass production using pressure die casting.
   - Any other suitable statement.

   Statements could include:

   Frosted acrylic is a good choice because the bathroom set must be durable enough to take everyday usage and wear-and-tear.
   The use of the chrome finished metal will blend in with most bathroom styles.

   Chrome plated metal is a good choice of material for set B because it does not corrode when in contact with water. It keeps its shiny appearance making it look both appealing and hygienic.
   - Resistant to shock *(Set A/B)*
   - Glass is tough/durable, can withstand weight

   Six valid statements at one mark each

   (c) • Justification of the types or manufacturing processes used in the production of the illustrated products and how production processes relate to the materials used.

   Acceptable processes:
   Set A – Injection moulding/compression moulding
   Set B – Die casting/Bending/Press-forming/Extrusion/Electro plating
   - How manufacturing/assembly techniques are influenced by volume of production.

   Statements could include:

   Standardisation of sizes, component parts all the same size.
   No further finishing required. Shapes suitable for process.
   Standardisation of components and materials chosen because they are easily sourced/formed.
   Suitable for mass/batch production – injection moulding, pressure die casting.
   Economy of scale – mass/continuous production.
   Any six relevant issues about materials, processes and their relationships regarding suitability.

   (d) Any four identified issues explained:
   - Fitness for its purpose
   - Durability to withstand continual use
   - Stability of free-standing units
   - Easy to clean/hygienic
   - Easy to use
   - Safety aspects of function
   - Maintenance issues
   - Easy to install
   - Any other acceptable issue.

   Four issues identified, four issues at one mark each

   (e) Any four identified issues described:
   - Cost (only if justified as a comparison)
   - Aesthetics
   - Durability
   - Brand name/image
   - Recycable
   - Hygienic
   - Look of a quality product
   - Compliment existing products
   - Any other acceptable issue.

   Statements could include:

   Set A
   'Set A' is made from a thermo plastic which makes it easy to clean and therefore hygienic

   Set B
   This bathroom set has a very traditional look because it is made from plated metal and has used similar features throughout the range. The set looks expensive and well made and should last a long time.
   Four issues at one mark each

   (f) Any four issues described in the context of ergonomics relating to bathroom accessory sets.

   Examples from:
   - Grip
   - Hand sizes (for access to components)
   - Finger traps
   - Surface texture to prevent slipping
   - Weight for lifting/strength issues
   - Colour - psychological
   - Feeling of cleanliness
   - Simple design/ease of use
   - Looks safe and simple to use *(Set A)*
   - Comfort

- Access for cleaning/maintenance
- Any other relevant answer.

Four statements at one mark each.

Total for Section A *30 marks*

SECTION B

2. **Annotated Sketches**
- Initial ideas/concepts
- Design development

Used to quickly communicate ideas/concepts without giving too much detail.

Working Drawings
- Planning for manufacture
- Production of prototypes

Used when dimensions/machining details are required allow accurate components to be manufactured.

Rendered 3D Computer Models
- Presenting a design proposal to the client
- Computer Testing/Evaluating
- Marketing

Statements could include:

A 3D computer model can be used to visually communicate a design proposal to a client. This will allow the client to make an initial assessment prior to production of a prototype.

Any suitable answer for each stage

One mark for identifying stage and one mark for each explanation

Total marks 6

3. (*a*) *Statements could include:*
- Increased use of landfill disposal due to limited lifespan
- Increased use of non renewable resources
- De-forestation
- Increased use of recycled/waste materials (re-use of standard components)
- Pollution produced during manufacture
- Any other relevant answer.

Two issues at one mark each

(*b*) *Statements could include:*

(i) **The consumer:**
- Assembly guidelines are clear (easy to assemble)
- Product is available in a variety of finishes
- Suited to modern designs
- Easy to transport
- Available in a wide variety of configurations
- Low cost
- Any other acceptable issue (must be specific not general).

(ii) **The manufacturer:**
- Less packing required
- Product diversification
- Use of CAD/CAM
- Standardisation of components
- Materials chosen because they are easily formed
- Jointing methods suitable for mass/batch production – spindle moulder, CNC, etc.
- Stock sizes suitable for flat pack production
- Low-skill requirements in production
- Manufacturing management systems such as JIT production
- Project planning of assembly
- Storage
- Any other justified answer.

(*c*) *Statements could include:*
- Consumer testing
- Use of standard components/simple knock down fittings
- Clear instruction leaflets
- Supplied tools
- Assembly components organised in a packet
- Any other justified answer.

To achieve full marks an extended answer is required

(*d*) *Statements could include:*
- High quality of materials
- High quality of finish achieved
- Greater durability than flat-pack
- Quality of look/attractive
- Build quality (dovetails etc)
- No assembly required
- Brand image
- Any other justified answer.

To achieve full marks an extended answer is required

Total marks 10

4. (*a*) *Statements could include:*
- Compression moulding is mostly used to make larger flat or moderately curved parts (suitable form for process)
- Is a high-volume, high-pressure method suitable for molding complex shapes
- One piece production
- Is one of the lowest cost molding methods
- Wastes relatively little material
- Accuracy with little shrinkage
- Thermosetting plastic process
- Any other relevant answer.

Two reasons at one mark each.

(*b*) **Material selection:**

Melamine Formaldehyde
Bakelite
Urea Formaldehyde
GRP
Phenol Formaldehyde

Justification:

Statements could include:
- Thermosetting plastic, electrical insulator, scratch resistance, strength/durability issues, temperature insulator.

One mark for identification of material
Two marks for justification

Total marks 5

5. (*a*) (i) **Initial concept**

Statements could include:
- Client needs
- Target market requirements
- Other products on the market
- Legislation restrictions
- IPR issues
- Knowledge of up-to-date materials and technology
- Appropriate materials for manufacture
- Any other justified answer.

One mark each valid point made in description. Two marks.

(ii) **Planning for Production**

Statements could include:
- Appropriate manufacturing techniques
- New manufacturing techniques available
- Production timings and costs
- Sub contractors/sources of standard components
- Delivery times (materials/standard components)
- Sources/quality of materials
- Any other justified answer.

One mark each valid point made in description. Two marks.

(*b*) **End user trials**

Statements could include:

- Safety issues
- Ergonomic issues relating to redesign
- Economic considerations (Pricing structure)
- Functional/Performance issues
- Environmental concerns/sustainability???
- Durability
- Variety of models/colours required
- Any other justified answer.

One mark each valid point made in description. Two marks.

Total marks 6

6. (*a*) *Statements could include:*

- Material in thin sections
- Material suited to process (malleable/ductile)
- Accuracy of sizes
- Repeatability/Consistency in quality
- Edges require no further finishing
- Increased strength of component after bending
- Any other relevant answer

One mark each valid point made in description.
Three marks.

(*b*) *Statements could include:*

- Ease of assembly
- Ease of replacement/repair
- Non permanent fixings allow for movement in the frame
- Quality issues
- Cheaper to purchase
- Saves time/Manufacturing costs
- Standard tooling
- Any other relevant answer.

One mark each valid point made in description.
Two marks.

(*c*) *Statements could include:*

- JIT
- Product diversification
- Lower costs
- Reduced need for customer/employee training
- Reduced storage/manufacturing space required
- Quality assurance procedures reduced
- Any other relevant answer.

One mark each valid point made in description.
Two marks.

Total marks 7

7. (*a*) *Statements could include:*

- Increased functionality
- Durability of materials
- Costs reduced due to improved manufacturing methods
- Improved reliability of operation
- Miniaturisation (portability)
- Fashion issues
- Planned obsolescence
- Technology transfer opportunities
- Environmental considerations
- More powerful/faster operation
- Colour/touch screen
- Multifunctional device
- Any other relevant answer.

One mark for basic explanation
Two issues explained. Two marks.

(*b*) *Statements could include:*

(i) **The designer**

- Ideas are the Intellectual Property (IP) of the creator, either of an individual or a company
- Ensure that they are not infringing other patents etc.
- Original design drawings must be kept safe as proof.
- An in-house designer has no rights to the design, all rights lie with his/her employer
- A freelance or consultant designer may be able to protect his design by use of a patent.

(ii) **The client/manufacturer**

- Will protect them by the use of patents and/or trademarks.
  but as patents are in the public domain these ideas may be copied by other companies in other parts of the world.
- Creative processes which generate new ideas may have commercial value
- Commercially valuable ideas can be at risk if not carefully protected, and others may gain commercial advantage as a result
- IP can have enormous commercial value and can be traded as a commodity.

One mark for basic explanation and two marks if the answer is extended to show deeper understanding of IPR issues

Total marks 6

Total for Section B *40 marks*

HIGHER PRODUCT DESIGN 2011

SECTION A

1. (*a*) The chair must:
   - be easily folded or unfolded
   - be stable (chair B)
   - supply adequate support when in use
   - be manufactured from durable materials that are suitable and appropriate for its function
   - be priced to suit the intended target market
   - ensure aesthetics suit the market niche or consumer aspirations
   - integrated cup holder (chair B)
   - production costs significantly less than selling price
   - be produced in a variety of colours to give target market a wider choice
   - be easy to clean/maintain
   - comply with relevant safety regulations
   - corrosion issues
   - must be portable
   - any other suitable statement.

   Six statements at one mark each

   (*b*) *Statements which identify issues such as:*
   - durability of material (*non corrosion*)
   - strength to weight issues
   - readily available materials
   - nylon – stretches to mould to body – dries quickly after rain
   - safety
   - suitability for production methods
   - function of component parts
   - aesthetic properties
   - ease of clean / hygiene
   - re-cycling
   - any other suitable statement.

   Statements could include:
   Aluminium offers an excellent strength to weight ratio which is ideal for use with Chair B as it has to be transported by the user.

   The nylon mesh offers an extremely light and hard wearing material and can be easily cleaned.

   Natural birch is a good choice of material for Chair A as it is durable, easily maintained and offers a good strength to weight ratio and when combined with the nylon material makes the product light yet robust.

   (*c*)
   - Identification of the types or manufacturing processes used in the production of the illustrated products and how production processes relate to the materials used.
   - Chair A – Spindle moulding, machine router, CNC.
   - Chair B – Injection moulding, Bending/forming, Extrusion.
   - How manufacturing/assembly techniques are influenced by volume of production.

   Statements could include:
   Standardisation of sizes, component parts all the same size.
   No further finishing required. Shapes suitable for process.
   Standardisation of components and materials chosen because they are easily sourced/formed.
   Suitable for mass/batch production – injection moulding.
   Economy of scale – mass/continuous production/JIT.

   One mark for correct identification of process to a maximum of three processes.

   (*d*) *Any four identified issues described:*
   - Fitness for its purpose
   - Durability to withstand continual use
   - Safety aspects of function
   - Maintenance issues (manufacture only – re tooling)
   - Quality of raw materials
   - Product testing
   - Well trained staff
   - Guarantees
   - Warranty
   - Quality of standard components
   - High quality finish
   - Any other acceptable issue.

   Four issues identified, four issues at one mark each

   (*e*) *Any identified niche market from:*
   - Camping
   - Holiday makers
   - Hill walkers
   - Climbers
   - Students
   - Anglers
   - Bird watchers
   - Festival goers
   - Any other acceptable answer.

   Statements could include:
   Chair A
   Chair A would be the ideal product for a hill walker as it is easily stored.

   Chair B
   This chair would appeal to the holiday maker as it can be easily transported in its folded form and is light yet durable and offers excellent support.

   One mark for identification plus one for each justification

   (*f*) *Any four issues described in the context of ergonomics:*

   Examples from:
   - Anthropometrics relating to seated position and back support
   - Hand sizes (for access to components)
   - Finger traps
   - Surface texture to prevent slipping
   - Weight for lifting/strength issues
   - Psychological issues – colour, ease of assembly, audible click when legs telescope
   - Comfort
   - Access for cleaning/maintenance
   - Any other relevant answer.

   Four statements at one mark each

SECTION B

2. (*a*) Piercing and blanking is suitable because:
   - Economies of scale
   - Repeatability
   - Accuracy
   - Shape of product
   - No finish required
   - Type of material used
   - Any other suitable answer.

   Two statements at one mark each

   (*b*) Pressing/Press Forming

   (*c*) **Suitable material:**
   - Stainless steel

   Justifications:
   - Corrosion resistance
   - Thin material

- Finish
- Aesthetics
- Hygiene
- Scratch resistance
- Chemical resistance

One mark for identification of material
One mark each for each justification

3. (*a*) *Any issues such as:*
- More opportunity for creativity
- More scope for innovation
- More opportunity to diversify into new related product ideas
- Opportunity for technological transfer.

One mark each description

(*b*) *Description including issues:*
- Identification of client requirements
- Financial constraints
- Key design issues
- Market requirements
- Target group
- Production volume
- Safety issues
- Market share/competition
- Brand image/aesthetics

One mark for each description

(*c*) *Any description that includes at least two issues from:*
- Ideas are the Intellectual Property (IP) of the company
- In-house designer has no IP rights
- IP can have enormous commercial value, and can be traded as a commodity
- Commercially valuable ideas can be at risk if not carefully protected
- Others may gain commercial advantage should designer leave company

One mark for each description

(*d*) There are five forms of protection
- Trademark
- Patent
- Registered Design
- Copyright
- Design Right

One mark for each description

(*e*) Laser sintering,
Fused deposition modelling

4. (*a*) *Any description that includes suitability such as:*
- Rotational moulding
- Thermoplastic process
- Hollow construction
- One piece construction
- Can dictate wall thickness
- Complex shapes can be formed
- No restriction on colour combinations/ addition of decals etc
- GRP
- Suitable for small batches
- Moulded in two halves and joined
- Strength issues
- Can be coloured separately (split colours)
- Customise finish
- Accept spray method of GRP

Example
The process of rotational moulding allows for one piece construction that makes the main body watertight.

One mark each description for both processes

(*b*) Any disadvantage relating to Rotational moulding
- High set-up costs
- Relatively long cycle times
- Choice of moulding material limited
- Powdered plastic rather than pellets required
- Some geometrical features difficult to mould
- Loading and unloading is labour intensive
- Any other suitable answer

One mark each description

5. Description should comment on the following:
- Reflects the landscape around it
- Shows natural wave shape
- Provides an open pathway to users
- Imitates Clydeside skyline
- Smooth lines combining with the strong geometric shapes
- Or any other appropriate points

Four statements at one mark each
Two marks awarded for extended answer

6. (*a*) *Any answer from:*
- Structured project planning of production (JIT)
- Increased quality assurance and control of production
- Increased productivity
- Reduction in stock wastage
- Less hours lost in production time
- Labour issues
- Manufacturing costs reduced
- Storage of component parts reduced
- Expertise of manufacture of bought-in components employed
- Reduced lead times

One mark for each description

(*b*) Description should include:
- Dependence on prompt delivery of components
- Component accuracy
- Quality assurance issues
- The company's requirement of outsourcing bought parts
- Bought parts may become obsolete
- Reliability of subcontractor
- Problems in meeting deadlines
- Any other justified answer.

Two issues at one mark each

(*c*) Description should include:
- Identification of alternative suppliers
- Identification of alternative components
- The ability to change suppliers
- Build in sufficient delivery timeslot
- Agreed quality assurance issues
- Any other justified answer.

Two issues

7. (*a*) Any justification that relates to:
- One piece construction
- Complexity of shape
- Strength issues
- Lighter in weight
- Colour combinations
- Maintenance.

One mark for each explanation

(*b*) Description should include:
- One piece design
- Streamlined / low profile design
- Ergonomic hand holds
- Ergonomic seating position
- Robust
- Any other justified answer.

Three issues three at one mark each

HIGHER PRODUCT DESIGN 2012

SECTION A

1. (*a*) The coat rack must (*any six from*):
    - must provide appropriate storage.
    - be accessible to the target users.
    - be capable of being securely fixed to the wall [CSU].
    - be easily moved [MCT].
    - be manufactured from durable materials that are suitable and appropriate for its function.
    - lifespan issues (e.g. methods of construction).
    - be priced to suit the intended target market.
    - ensure aesthetics suit the market niche or consumer aspirations.
    - production costs significantly less than selling price.
    - be produced in a variety of colours to give target market a wider choice.
    - be easy to clean/ maintain.
    - conform to appropriate safety regulations.
    - any other suitable statement.

    (*b*) *Any six statements which identify issues such as:*
    - durability of material *(non corrosion)*.
    - readily available materials (e.g. standard forms of supply).
    - strength to weight issues.
    - safety.
    - suitability for production methods.
    - function of component parts.
    - aesthetic properties.
    - ease of clean / hygiene.
    - re-cycling (metal components only).
    - any other suitable statement.

    (*c*)
    - Justification of the types or manufacturing processes used in the production of the illustrated products and how production processes relate to the materials used.

      MCT – Spindle moulding, machine router, bending & forming, extrusion, CNC

      CSU – Injection moulding, Blow moulding, Extrusion.
    - How manufacturing/assembly techniques are influenced by volume of production.

    Statement could include:

    Standardization of components/ sizes, component parts all the same size. no further finishing required. Process related to shape of component.

    Repeatability and accuracy. Economies of scale. JIT.

    Any six relevant issues about materials, processes and their relationships regarding suitability.

    (*d*) *Any four identified issues described from:*
    - items safely stored.
    - durability to withstand continual use.
    - stability of free-standing units .
    - easy to clean/hygienic.
    - easy to use.
    - safety aspects of function (e.g. accessibility).
    - safety aspects relating to production (e.g. noise, dust, safety equipment).
    - maintenance issues.
    - any other acceptable issue.

    (*e*) *Explain any four identified functional aspect from:*
    - storage.
    - accessibility.
    - space saving.
    - ease of assembly/ fixing.
    - mobility [MCT].
    - stability.
    - durability.

    Any identified niche market from:
      - Nursery school.
      - Primary school.
    - Any other acceptable issue.

    (*f*) *Explain any four issues from:*
    - anthropometrics relating to adult and child users.
    - finger traps.
    - surface texture to prevent slipping.
    - weight for moving/strength issues.
    - colour – aesthetic appeal.
    - comfort.
    - access for cleaning/maintenance.
    - security.
    - any other relevant answer.

SECTION B

2. (*a*) Suitable material (*any one from*):
    - High Density Polyethylene.
    - Low Density Polyethylene.
    - Polypropylene.
    - PVC.
    - Polyethylene terephthalate (PET).
    - or other suitable material.

    Reason (*any one from*):
    - easy to mould.
    - durability.
    - easy to clean.
    - chemical resistance.
    - any other acceptable answer.

    (*b*) *Any two features identified from:*
    - hollow part.
    - split lines.
    - ejection mark.
    - visual signs of parison crimping.
    - good external surface detail.

    (*c*) Product needs to be lightweight.

3. **Scale Models**

    Used to quickly communicate ideas/concepts without giving too much detail.
    - Checking ergonomic aspects.
    - Developing concept ideas.
    - Developing aesthetic aspects.
    - Aspects of testing.
    - Feedback to client.
    - Any other acceptable answer.

    Test Models

    Used when determining functionality to allow accurate components to be manufactured.
    - Determining structural suitability.
    - Health & safety compliance.
    - Gauging functional efficiency.
    - Material properties.
    - Any other acceptable answer.

    Prototypes

    Used to show the final design to simulate the final design, aesthetics, materials and functionality of the intended design.
    - Check for any flaws.
    - Test efficiency.
    - Performance issues.
    - Check public opinion.
    - Any other acceptable answer.

    Two marks for each model type.

4. (*a*) *Explanation should include at least two issues from:*
- Exclusivity of product/company.
- Legal rights.
- Company branding.
- Prevent illegal use of logo/name.
- Or other appropriate reason.

(*b*) *Explanation should include at least two issues from:*
- Register ownership.
- Protect company rights.
- Prevent idea/patent being copied.
- Can lease design.
- Or other appropriate reason.

5. (*a*) Anthropometric issues (*any three from*):
- Chair seat width.
- Popliteal height increased to clear stairs.
- Longer back length.
- Adjustability of head restraint.
- Belt size.
- Hand grip.
- Or other suitable answer.

(*b*) Physiological issues (*any three from*):
- Weight of chair.
- Strength of assistant to weight of user ratio.
- Comfort of user.
- Strain on back of assistant.
- Lever lengths.
- Ease of folding/unfolding.
- Weight of passenger.
- Or other suitable answer.

(*c*) Psychological issues (*any three from*):
- Confidence issues surrounding user.
- Stability of chair.
- Belt for safety.
- High resolution colours used.
- Construction looks robust.
- Simplicity of use.
- Or other suitable answer.

6. (*a*) *Justification should include at least two issues from:*
- Component size.
- One piece construction.
- Suitable for small production rates/Economies of scale/Batch.
- Fairly complex shapes can be production.
- Highly skilled workers needed.
- Low equipment costs.
- Shape of product (no undercuts).
- Any other suitable answer.

(*b*) *Explanation should include at least two issues from:*
- Sand casting produces surface defects.
- Sand indentation.
- Removal of excess material (Runner/Riser/Flash etc).
- Limited quality control.
- Does not allow close tolerances e.g. wheel nut spacing.
- Does not give a smooth surface finish.
- Any other suitable answer.

(*c*) *Comparison should include three issues from:*
- Pressure die casting
- Highly complex shapes can be produced.
- High quality repeatability.
- Accuracy.
- Consistent quality.
- No further finishing required.
- Thin walls can be produced reducing weight.
- Flexibility of production.
- Speed of process.
- Reduced storage required for patterns.

(Two marks awarded for extended answer)

(*d*) *Explanation should include two issues from:*
- Improved mechanical strength.
- Improved heat resistance.
- Improved chemical resistance.
- Improved aesthetic appearance.
- Improved performance issues.
- Any other appropriate answer.

7. (*a*) Factors considered (*any two from*):
- Materials (e.g. performance/ durability).
- Functionality (e.g. sports).
- Market niche.
- Safety (e.g. UV protection).
- Fashion.
- Cost.
- Or other appropriate answer.

(*b*) *Explanation should give indication in differences between fashion and style.*

Fashion, a general term for a currently popular style or practice.
- Current trend.
- Determined by season.
- Temporary.

Style, a distinctive and identifiable appearance.
- Long lasting appeal.
- Quality product (Branding).
- Or any other appropriate answer.

One mark for each appropriate point made

HIGHER PRODUCT DESIGN 2013

SECTION A

1. (*a*) The Lawnmower must (*any six from:*)
    - cut grass
    - have an adjustable height of cut
    - be easily stored when not in use
    - be easily manoeuvred
    - be manufactured from durable materials that are suitable and appropriate for their function
    - be priced to suit the intended target market
    - ensure aesthetics suit the market niche or consumer aspirations
    - be used in larger gardens (lawnmower B)
    - cut on sloping ground (lawnmower A)
    - have production costs significantly less than selling price
    - look obvious to use
    - be easy to empty (lawnmower B)
    - pick up cut grass (lawnmower B)
    - be easy to clean/maintain
    - be safe (when in context) eg blades guarded
    - have reference to suitable cable length (lawnmower A)
    - comply with relevant safety regulations
    - any other suitable statement.

   (*b*) Statements which justify issues such as:
    - durability of material/impact resistance (*non corrosion*)
    - strength to weight issues
    - readily available materials
    - Nylon – self lubricating gears
    - HSS/tool steel – hardness
    - rubber – excellent grip
    - chemical resistance
    - comfort of foam
    - foam gives excellent grip
    - suitability for production methods
    - function of component parts
    - aesthetic properties
    - ease of cleaning
    - re-cycling
    - any other suitable statement.

   Aluminium offers an excellent strength to weight ratio which is ideal for use with lawnmower B as it has to be transported by the user.

   The nylon mesh offers an extremely light and hard wearing material that can be easily cleaned.

   PP is a good choice of material for lawnmower A as it is durable, easily maintained and offers a good strength to weight ratio.

   Metal alloy is durable.

   (*c*) *Suitable processes:*

   Lawnmower A – Extrusion, Bending, Injection Moulding, Piercing and Blanking, Plastic/Powder/ Dip coating, Sharpening

   Lawnmower B – Injection moulding, Compression Moulding, Welding, CNC Machining, Bending/press forming, Extrusion, Pressure die casting, sharpening, spray painting.

    - How manufacturing/assembly techniques are influenced by volume of production
    - Process is suitable for the material eg (sheet metal for press forming)

   Statements could include:

   Standardisation of sizes, component parts all the same size. No further finishing required. Shapes suitable for process.

   Standardisation of components and materials chosen because they are easily sourced/formed.

   Suitable for mass/batch production – injection moulding. Economy of scale – mass/continuous production/JIT

   (*d*) *Any four issues explained in the context of ergonomics with examples from:*
    - Anthropometrics relating to handle grip, position and locking
    - Position of handles for operation (dead man's switch)
    - Handle adjustment (Height)
    - Hand size (for access to components)
    - Finger traps
    - Surface texture to prevent slipping
    - Weight for lifting/strength issues
    - Strain/fatigue issues
    - Psychological issues – colour, ease of assembly, looks safe and easy to use
    - Comfort during use
    - Access for cleaning/maintenance
    - Any other relevant answer.

   (*e*) *Any four descriptions from:*
    - Cost (only if compared)
    - Aesthetics
    - Durability
    - Brand name/Image
    - Recycling
    - Easy to use/store
    - Lightweight
    - Grass catcher (Lawnmower B)
    - Safety
    - No need of electricity (Lawnmower B)
    - Easy to move due to wheels (Lawnmower B)
    - lower disposable income (Lawnmower A)
    - No cable, so it can be used more than 20m from a power point (Lawnmower B)
    - Can be used on severe slopes (Lawnmower A)
    - Any other acceptable answer.

   (*f*) *Any four identified issues described from:*
    - Fitness for its purpose (*not simply cut grass*)
    - Durability to withstand continual use/outdoor environment
    - Safety aspects of function – dead man's trigger; electrical safety
    - Maintenance issues
    - Ease of use
    - Easy to clean
    - Ease of adjustment
    - Choice of materials (relating to function)
    - Stability of lawnmower
    - Portability
    - Storage
    - Any other acceptable issues.

SECTION B

2. (*a*) The benefits of drop forging are:
    - Stronger than similar cast or machined products/improved strength characteristics
    - Repeatable process
    - Accuracy
    - Complex shape
    - Good surface detail
    - One piece construction
    - Or any other suitable answer.

(*b*) The features of drop forging are:
- Parting lines/flashing/flash removal/split lines
- Quality of surface detail/texture
- One piece construction
- Contrast of finishes
- Evidence of further finishing might be apparent
- Draft angle/relief profile
- Or any other suitable answer.

(*c*) While most metals can be dropped forged, the material selected should be appropriate to the product.
- Chromium alloy steel
- High carbon steel
- Nickel/Nickel alloys
- Titanium

Reason for choice:
- Durability
- Toughness
- Resistant to corrosion (*Chromium alloy steel, stainless steel, Nickel/Nickel alloys, Titanium*)
- Any other suitable answer.

3. *Description should include at least two aspects from:*
- Aesthetics to enhance function
- Aesthetics to promote/enhance style
- Shape/Layout of buttons enhanced by aesthetic
- Form of product
- Colour contrasts/harmonies
- Contrast/harmonies in shapes
- Gloss surface compared with matt surface
- Texture
- Any other suitable answer.

4. (*a*) *Any two issues explained with reference to the product:*
- Focussing product to specific target group
- Aspirational group – prestige, brand awareness
- Financial aspects
- Profitable segment of a market
- Satisfying specific market needs
- Any other relevant answer.

(*b*) *Any two issues described from:*
- Fewer types of materials per product
- Lower volume of materials per product
- Materials that are easy to recycle/renewable
- Materials easily identifiable
- Use alternative materials
- Product easily dismantled for recycling
- Any other relevant answer.

(*c*) *Any two issues described from:*
- Demand for product
- Price point (Usually towards top end of market for batch produced items)
- Flexibility of design (batch)
- Flexibility of production
- Less machinery required but generally multipurpose machines to allow retooling (batch)
- Particular components which must be batch produced
- Standard components
- Production planning
- Production costs
- Any other relevant answer.

5. (*a*) *Any two from:*
- Use of Rapid Prototyping to shorten Research and Development time
- Outsourcing to specialists
- Ensure reliable delivery of raw materials
- Efficient production and process scheduling
- Reduce the number of processes employed
- Reduce transit time
- Efficient Quality Assurance procedures
- Higher staffing level/work longer hours
- Reduced Research and Development time
- Any other relevant answer.

(*b*) Tablet 2 is the most viable.

Reasons for choice:
- Although the sales are lower in terms of volume the profits are higher (from graph)
- Longer and costlier lead time for Tablet 1
- Less negative profit during lead in for Tablet 2
- Sales start earlier (Tablet 2)
- Sales last longer (Tablet 2)
- Tablet 2 in profit longer.

(*c*) *Any two from:*
- Reduce price of product
- Special offers (eg Free tablet case)
- Increased advertising
- Updated versions (software)
- Introduce special editions/additional features
- Any other relevant answer

6. (*a*) *Any two from:*
- Cost of manufacture reduced (simple tooling)
- Speed and simplicity of assembly
- Can be purchased as standard components
- Unskilled labour can be employed
- Can be Flat-packed/Easily Stored/Easily Transported
- Fittings are produced to suit manufactured boards
- Standard components/parts
- Speed of manufacture is increased
- Any other relevant answer.

(*b*) *Any two from:*
- Quality of finish
- Aesthetics
- Fashion/style
- Durability
- Strength issues
- Recyclability
- Any other relevant answer.

(*c*) *Any two from:*
- Ease of recycling
- Changes in fashion/style
- Ability to change doors/worktops
- Durability of carcase materials and fittings
- Maintenance/replacement issues
- Obsolescence of appliances
- Any other relevant answer.

7. (*a*) *Explanation should include any two from:*
- Saving money on product development phase
- Reduced labour costs
- Money saved on testing components
- Quicker development phase
- Faster resetting time for tests
- Quicker feedback
- Exact conditions can be replicated
- Material wastage
- No human error
- Can run 24/7
- Safety aspects
- Or any other suitable answer.

(*b*) (i) Suitable materials;
- Polycarbonate/ABS mix
- ABS
- Polycarbonate
- SAN (Styrene Acrylonitrile Resin)
- Or any other suitable material

(ii) Advantages:
- No material is lost
- High strength
- Cost effective
- Component accuracy/detail
- Reduced lead times
- Assembled prototypes produced
- Functional prototypes/products
- Multiple materials can be used
- Robust/tough
- Impact resistance
- Choice of colours
- Or any other acceptable answer.

Disadvantages:
- Can show "ribbing" from layers
- High set up costs
- Slow compared to other RP processes
- Supports in structure might be needed
- Large areas require longer build times
- Or any other suitable answer.

8. (*a*)
- Observation
- Questionnaires/surveys
- User group testing
- Information from experts/focus groups
- Any other suitable description.

(*b*) *Any three from:*
- Strength issues
- Fatigue
- Dexterity
- Mobility
- Any other suitable description.

(*c*) *Any two from:*
- Morphological Analysis
- Brainstorming
- Storyboard/mood board/image board
- Technology transfer
- Mind mapping/spider diagram
- Analysis of existing products
- Lateral thinking
- Any other suitable answer.

(*d*)
- Use of programmes such as powerpoint
- Manual graphics
- Computer graphics
- Animation
- Modelling/prototyping
- Demonstration
- Any other suitable answer.